THERIOGENOLOGY GURU

INDEX

PAGE	SUBJECT	PAGE	SUBJECT

Made in the USA
Las Vegas, NV
19 December 2024

14943796R00059